STEAM IN THE 1950s

STEAM IN THE 1950s

THE RAILWAY PHOTOGRAPHS OF ROBERT BUTTERFIELD

COMPILED BY BRIAN J. DICKSON

The
History
Press

First published 2020

The History Press
97 St George's Place,
Cheltenham GL50 3QB
www.thehistorypress.co.uk

British Library Cataloguing in Publication Data.
A catalogue record for this book is available from the British Library.

ISBN 978 0 7509 9370 8

Typesetting and origination by The History Press
Printed in Turkey by Imak

Front Cover: Sunday, 1 September 1957. Ex-LMS Class 8P Princess Coronation 4-6-2 No. 46238 *City of Carlisle* is seen passing Hest Bank station at the head of an 'up' express. Constructed at Crewe Works during 1939 in a streamlined form, the streamlining being removed in 1946, she would initially be allocated to Camden shed. From 1948 until she was withdrawn during 1964 she would be based at Carlisle Upperby shed.

Back Cover: Wednesday, 22 June 1955. This busy scene at St Margarets shed in Edinburgh shows a number of ex-NBR locomotives standing over ash pit roads at the back of the shed. Ex-NBR Class J Superheated Scott (LNER Class D30) 4-4-0 No. 62427 *Dumbiedykes* was constructed at Cowlairs Works in 1914 and numbered 418 by the NBR. Later becoming No. 9418 and then 2427 with the LNER, she would be withdrawn from service during 1959. The character 'Laird o' Dumbiedykes' is prominent in the 1818 published Scott novel *The Heart of Midlothian*.

Half Title Page: Sunday, 15 April 1951. Seen passing Preston is The Royal Scot heading north in the hands of ex-LMS Class 8P Princess Coronation 4-6-2 No. 46254 *City of Stoke-on-Trent*. She was a product of Crewe Works in 1946 and was withdrawn in 1964.

Title Page: Thursday, 23 June 1955. Standing outside the shed at Mallaig is ex-LNER Class K1 2-6-0 No. 61997 *MacCailin Mor*. Originally constructed at Darlington Works in 1939 to a design by Nigel Gresley as a three-cylinder Class K4 locomotive, she would be numbered 3445 by the LNER. She was rebuilt during the Edward Thompson regime at Doncaster in 1945 as a two-cylinder precursor for a proposed class of similar locomotives to be classified K1. Eventually, seventy examples of the Class K1 were constructed by the NBL in Glasgow during the Arthur Peppercorn period in 1949–50. *MacCailin Mor* would be withdrawn from service in 1961.

INTRODUCTION

Robert Butterfield had a lifelong passion for railways. He devoted his career to working for British Railways; was a dedicated enthusiast, photographer and railway modeller; and in retirement spent many happy hours chasing steam specials, particularly on the Carlisle to Settle line.

Born in 1932 in Shipley near Bradford, Robert's interest in railways began at an early age when he started trainspotting at Hest Bank station as a schoolboy and this led him to join the railways. The trainspotting bug remained with him for the rest of his life and he became an ardent enthusiast of the former LM&SR, particularly the locomotives of the Duchess (Princess Coronation), Royal Scot and Jubilee classes. He was a lifelong member of the Railway Correspondence and Travel Society (RCTS) and took part in numerous trips, shed bashes and works visits organised during the 1950s and '60s. He also travelled extensively using his allocation of 'free passes' and enjoyed making epic journeys to explore the Scottish network.

His professional career started in 1950 when, at the age of seventeen, he joined British Railways as a clerk at Frizinghall station in Bradford. This was swiftly followed by National Service in the RAF from 1951 to 1953, being posted to Wildenrath, Germany, and working in Air Traffic Control as an aerodrome assistant in the control tower. He returned to the railway in early 1954 and successfully completed the station-master training course at Derby. He then worked at various locations in the North Eastern Region and was based at Redcar for six months. His first major post was in the late 1950s to become station master on the Calder Valley Line at Luddendenfoot station (where Branwell Bronte worked as 'clerk in charge' in the early 1840s). This station had a large goods yard serving a busy coal drop and the many wool mills in the area, but would be closed to passengers in September 1962.

In the early 1960s Robert became station master at Longwood and Milnsbridge and Goldcar stations on the Standedge Line, and lived in a railway house overlooking the magnificent viaduct at Milnsbridge. With the closure of these stations he was forced to change direction and by the early 1970s was working as a relief booking office manager covering many stations in West Yorkshire. In the late 1970s he worked in the offices at Healey Mills Marshalling Yard for several years (surviving an armed raid on the payroll in a neighbouring room), before spending the remainder of his career in the booking office at Huddersfield station. He finally retired from what had become Regional Railways in April 1993 having given forty-three years of service to the railway.

Robert Butterfield was an accomplished photographer and during National Service he purchased an expensive-made Agfa Record 1 F1:4.5 105mm lens 6 x 9 folding camera that used 120 film and in the late 1960s purchased a Halina Paulette F1:2.8 45mm 35mm camera and used colour transparency film. By the 1980s and 1990s he tended to do most of his photography using a video camcorder.

The photographs in this book, which are laid out chronologically, show how widely he travelled in the London Midland, Eastern, North Eastern and Scottish Regions. Visits to the Western and Southern Regions were very rare and very few of his photographs from these regions are dated within the collection and are therefore not included. Scotland was a favourite destination with visits to that country recorded every year from 1952 until 1956. On these many tours he managed to photograph many of the pre-grouping classes still in use, ex-NBR 4-4-0 and 0-6-0s, ex-Caley 'Standard Passenger' tanks and most of the graceful ex-GNoSR 4-4-0s still in operation. Another favoured location was Hest Bank with its water troughs lying just north of the station, this provided opportunities for many excellent photographs of his favourite classes – Princess Coronations, Royal Scots and Jubilees.

Saltaire and its magnificent mill buildings and the line just to the west of the station also became favoured spots.

After the demise of main-line steam in 1968, he served as Chairman of the Bradford Trolleybus Association for many years and spent several years filming the trolleybuses of Bradford before the closure of the system in 1972. By the late 1970s he became interested in seeing and photographing Royal Mail Postbuses in the remote Scottish Highlands and in 1979 purchased a retired Commer Postbus from Lockerbie for conversion to a dormobile for family holidays. In the 1980s and '90s Robert took advantage of his free travel passes to travel overnight by train to Innsbruck in Austria, Garmisch-Partenkirchen in Germany and the Gotthard Pass in Switzerland where he enjoyed filming the local transport systems.

As a keen railway modeller he constructed a 00-gauge model railway based on the Carlisle to Settle line and hand built a superb set of LM&SR coaches, which were fitted with sprung suspension, carefully lined with a bow pen and adorned with Thames-Clyde Express roof boards. In retirement he continued to focus on his love of railways and purchased a 3½in-gauge live-steam Stanier Mogul, attending the Brighouse Model Engineers Club to run the locomotive. This was not without its risks – on one occasion the front pony truck derailed on a trestle viaduct section of track and on another he had to hastily drop the fire after the injectors stopped working.

Robert was married in 1959 to Betty, the daughter of a Manningham driver, and he died in 2001 at the relatively early age of 68. His two sons Michael and Richard have both inherited his love of railways.

October 1952. Seen departing from Carlisle at the head of a Glasgow to Manchester express is BR Standard Class 6P5F 4-6-2 No. 72006 *Clan Mackenzie* piloting ex-LMS Class 6P5F Jubilee 4-6-0 No. 45645 *Collingwood*. The 'Clan' was a product of Crewe Works during 1952 that would only give fourteen years of service before being withdrawn in 1966. The Jubilee was also a Crewe Works product that entered service during 1934 and would be withdrawn in 1963.

October 1952. Also seen departing from Carlisle is ex-LNER Class A3 4-6-2 No. 60101 *Cicero* at the head of an Edinburgh to London St Pancras express. Named after the 1905 winner of the Derby, she was constructed at Doncaster Works during 1930 and would receive a double chimney in 1959. Allocated solely to Scottish sheds during her working life, she would be withdrawn in 1963.

Opposite top: It is approaching noon on this day in October 1952 and at the south end of Carlisle station ex-LMS Class 8P Princess Coronation 4-6-2 No. 46239 *City of Chester* waits to relieve the locomotive working the 'up' Royal Scot. Standing adjacent is ex-LMS Class 6P5F Jubilee 4-6-0 No. 45605 *Cyprus* waiting to pilot the train engine working the 'up' Thames-Clyde Express. No. 46239 was constructed at Crewe Works during 1939 in a streamlined form, which would be removed in 1947. No. 45605 was a product of Crewe in 1935. Both locomotives would be withdrawn from service during 1964.

Opposite bottom: Jubilee No. 45605 *Cyprus* is seen here departing from Carlisle station as pilot to ex-LMS Class 7P Royal Scot 4-6-0 No. 46109 *Royal Engineer*, which is working the 'up' Thames-Clyde Express. The 'Scot' was a product of the NBL during 1927 that would be rebuilt with a taper boiler in 1943 and be withdrawn in 1962.

Right: October 1952. Waiting to depart from Mallaig station with a working to Fort William is ex-GNR Class H3 (LNER Class K2) 2-6-0 No. 61781 *Loch Morar*. Constructed by Kitson & Co. during 1921, she would be allocated to the Scottish area of the LNER during 1933 to work the West Highland Route and was named at that time. She would be withdrawn in 1958.

Opposite top: October 1952. This wonderful scene taken from the carriage of a Mallaig to Fort William working shows ex-GNR Class H3 (LNER Class K2) 2-6-0 No. 61781 *Loch Morar* crossing the Glenfinnan Viaduct. The twenty-one span viaduct was constructed by Robert McAlpine & Sons using concrete, earning the company head the nickname 'Concrete Bob'. Completed during 1899, it crosses the River Finnan close to the head of Loch Sheil, the line between Fort William and Mallaig being opened in 1901.

Opposite bottom: October 1952. At Carlisle Citadel station, ex-NBR Class C (LNER Class J36) 0-6-0 No. 65312 is looking in good clean external condition with the tender bearing the early British Railways identity. Constructed at Cowlairs Works in 1899 and numbered 754 by the NBR, she would become 9754 and later 5312 with the LNER. She would be withdrawn after sixty-three years of service during 1962.

Above: October 1952. At Fort William ex-NBR Class C (LNER Class J36) 0-6-0 No. 65237 is shunting some coal wagons. Designed by Matthew Holmes and entering service from 1888, over 160 examples of the class were delivered from Neilson & Co. and Sharp Stewart & Co. in addition to the NBR's own works at Cowlairs. The example seen here was from the works of Neilson & Co., entering service during 1891. Numbered 663 by the NBR, she would become 9663 and later 5237 with the LNER and be rebuilt in 1916. She would be withdrawn in 1962 having served for seventy-one years.

October 1952. Working as station pilot at Oban, ex-CR Class 439 (LMS Class 2P) 0-4-4 tank No. 55196 has paused between duties. Constructed at St Rollox Works in 1909, she would be with withdrawn in 1955. The Caley 0-4-4 tank had its origins with the Dugald Drummond Class 171 introduced in 1884. Later Locomotive Superintendents Hugh Smellie, John McIntosh and William Pickersgill all introduced versions with varying cylinder sizes and other detail changes using the general description 'Standard Passenger Tank'. The LMS power classification covered both 1P and 2P versions. No. 55196 had been constructed during the John McIntosh period and is a Class 439 locomotive.

June 1953. With the stunning background of the Sir Titus Salt Textile Mills at Saltaire, ex-LMS Class 4P Compound 4-4-0 No. 41103 is seen piloting ex-LMS Class 7P Royal Scot 4-6-2 No. 46117 *Welsh Guardsman* at the head of the 'up' Thames-Clyde Express. The Compound was constructed at Derby Works during 1925 and would be withdrawn from service in 1957 with the Scot coming into service from the NBL in Glasgow during 1927. She would be rebuilt with a taper boiler in 1943 and be withdrawn in 1962.

Opposite top: June 1953. At Manningham shed near Bradford, William Stanier-designed ex-LMS Class 3P 2-6-2 tank No. 40178 is waiting for its next duty. Constructed at Derby Works in 1938 and numbered 178 by the LMS, it would be withdrawn during 1961 having given only twenty-three years of service.

Opposite bottom: June 1953. Seen passing Hirst Wood, to the west of Shipley, ex-LMS Class 4P 'Compound' 4-4-0 No. 41100 is working a westbound mixed train which contains a six-wheeled tank wagon. The locomotive is carrying 'Empty Coaching Stock' (ECS) lamps and is sporting a 20A Leeds Holbeck shed code. Entering service from Derby Works during 1925 and numbered 1100 by the LMS, she would be withdrawn in 1959. In the background can be glimpsed the prominent tower of the Saltaire United Reform Church, which was constructed in 1859.

Above: June 1953. Waiting to exit from Manningham shed in the late evening sunshine, a crew member is telephoning the signal box. Ex-L&YR Class 5 2-4-2 tank No. 50686 was a product of Horwich Works during 1892. Constructed to a design by John Aspinall, it would serve for sixty-three years before being withdrawn during 1955.

Left: Friday, 19 June 1953. Near Leeds Holbeck, ex-LMS Class 4F 0-6-0 No. 44570 makes a spectacular start whilst managing some shunting. Note the shunter's pole on the front running board. A product of Crewe Works in 1937, she would be withdrawn during 1965.

Opposite top: Sunday, 21 June 1953. Standing in a packed St Margarets shed yard in Edinburgh is ex-LNER Class V3 2-6-2 tank No. 67606. Originally constructed as a V1 locomotive at Doncaster Works in 1930 and numbered 2906 by the LNER, she would later become No. 7606 and be rebuilt as a V3 in 1952. She would spend the bulk of her working life based at St Margarets and be withdrawn in 1962.

Opposite bottom: Sunday, 21 June 1953. Seafield shed in Edinburgh was more of a large open yard where locomotives could be serviced and stabled, and was situated to the eastern side of Leith Docks. It provided locomotives for shunting in the dock area and the surrounding yards. Waiting for the start of the working week and at the head of this row of locomotives is ex-NBR Class G (LNER Class Y9) 0-4-0 saddle tank No. 68115. Constructed at Cowlairs Works during 1897, she would give sixty years of service before being withdrawn in 1957.

This spread shows three examples of the William Reid-designed powerful Class B 0-6-0s introduced during 1906 with a total of seventy-six examples entering NBR service, the last in 1913. Initially all were equipped with saturated steam boilers, but over a period of many years they were all rebuilt with superheating boilers. Classified J35 by the LNER the class had a long lifespan with the last being withdrawn in 1962.

Above: Sunday, 21 June 1953. Seen standing at Dunfermline shed is No. 64493. She was the product of the NBL Queen's Park Works in 1909 and would be numbered 206 by the NBR. Becoming 9206 and later 4493 with the LNER, she was rebuilt with a superheating boiler in 1933 and would be withdrawn from service in 1960.

Opposite top: Monday, 22 June 1953. With a busy Forfar locomotive shed in the background, No. 64468 is seen departing from Forfar station with a local working to Arbroath. Constructed during 1906 by the NBL Atlas Works, she would be numbered 856 with the NBR, becoming No. 9856 and later 4468 with the LNER. She was rebuilt with a superheating boiler in 1925 and withdrawn during 1960.

Opposite bottom: Monday, 22 June 1953. Standing at Forfar shed is No. 64520, which was constructed at Cowlairs Works during 1910 and numbered 120 by the NBR. She would become No. 9120 and later 4520 with the LNER and be rebuilt with a superheating boiler during 1933. She would be withdrawn in 1959.

Above: Monday, 22 June 1953. In Kittybrewster shed yard standing over the ash pit is ex-NBR Class K (LNER Class D34) 'Glen' 4-4-0 No. 62479 *Glen Sheil*. A product of Cowlairs Works during 1917, she was numbered 298 by the NBR and would become 9298 and later 2479 with the LNER. The 'Glens' were the reliable workhorses on the West Highland line for many years until after the grouping when some K2 2-6-0's were allocated to the route. *Glen Sheil* would end her days allocated to Kittybrewster shed and be withdrawn during 1961.

Opposite top: Monday, 22 June 1953. Bearing a 61A shed code, ex-GNR Class H3 (LNER Class K2) 2-6-0 No. 61793 is standing at its home shed, Kittybrewster. Constructed by Kitson & Co. during 1921 she would be numbered 4703 and later 1793 by the LNER. She would remain one of the unnamed Scottish members of the class and be withdrawn in 1959.

Opposite bottom: Monday, 22 June 1953. Standing beside the coaling stage at Forfar shed is ex-LMS Class 5MT 2-6-0 No. 42800. Constructed at Crewe Works in 1928 and numbered 13100 by the LMS, she would later be numbered 2800. Ten examples of the class were allocated to ex-Highland Railway territory by the LMS during 1928 but were restricted to working the main line between Perth and Inverness via Slochd. Spending all her working life allocated to Scottish sheds, she would be withdrawn during 1965.

Monday, 22 June 1953. In ex-works condition, ex-NBR Class S (LNER Class J37) 0-6-0 No. 64618 had been constructed by the NBL Atlas Works during 1920 and numbered 84 by the NBR. Later becoming 9084 then 4618 with the LNER, she would be withdrawn from service in 1966. Seen at Kittybrewster shed in Aberdeen, note the poor quality of coal in the tender.

Monday, 22 June 1953. Picking up speed at the head of a Glasgow-bound express near Ferryhill, Aberdeen, is BR Standard Class 5MT 4-6-0 No. 73007. Bearing a 63A Perth shed code and fitted with a small snowplough, she was constructed at Derby Works in 1951 and allocated to Perth. She would be withdrawn after only fifteen years of service during 1966. The Standard 5s were the very able successors to the Stanier 'Black 5s' with 172 examples being constructed at both Derby and Doncaster Works over a period of six years from 1951 to 1957.

Tuesday, 23 June 1953. Seen near Nigg Bay south of Aberdeen, Class A2 4-6-2 No. 60531 Bahram is at the head of the 'up Aberdonian' comprising of fifteen coaches with through carriages and sleeping cars to London King's Cross. Constructed at Doncaster Works in 1948, she was allocated to Ferryhill shed in Aberdeen for the bulk of her working life and would be withdrawn from service during 1962.

Tuesday, 23 June 1953. Waiting to depart from Kittybrewster shed is ex-CR Class 439 (LMS Class 2P) 0-4-4 tank No. 55185. Entering service during 1907 from St Rollox Works, she would be numbered 466 by the Caley and become 15185 with the LMS. She was allocated to Keith shed during 1952 and would be withdrawn from service in 1961.

Tuesday, 23 June 1953. Maud Junction was the starting point for the sixteen-mile branch to Fraserburgh which was opened during April 1865, the Dyce to Peterhead line having been opened three years earlier in July 1862. Although fish from both ports could be said to constitute the bulk of the goods traffic, meat (primarily beef) being carried to the large markets in Aberdeen was also a good source of revenue. Here BR Standard 4MT 2-6-4 tank No. 80021, with a rake of cattle wagons in tow, has stopped at the north end of the station on the Fraserburgh branch to take water. Constructed at Brighton Works in 1951, she would be withdrawn during 1964.

Tuesday, 23 June 1953. Having been turned on the turntable at Fraserburgh shed, BR Standard Class 4MT 2-6-4 tank No. 80004 simmers between duties. Constructed at Derby Works, she entered service during 1952 and was withdrawn in 1967. Note the impressive granite stonework of the Fraserburgh South Parish Church in the background, which was built in 1878, a full thirteen years after the railway opened in 1865.

Tuesday, 23 June 1953. Ex-GNoSR Class F (LNER Class D40) No. 62278 Hatton Castle is standing at Fraserburgh shed showing the graceful curve of the splashers, the slender, slightly tapering chimney and the tall dome of this Thomas Heywood version of the earlier William Pickersgill Class V design. Constructed by the NBL during 1920, she would be numbered 50 by the GNoSR and become No. 6850 and later 2278 with the LNER and be withdrawn from service in 1955.

Above: Tuesday, 23 June 1953. It is generally accepted that the Great North of Scotland Railway built some of the most graceful 4-4-0 locomotives ever to work in Great Britain. Lending credence to that view is this photograph of ex-GNoSR Class F (LNER Class D40) 4-4-0 No. 62276 *Andrew Bain* standing at the south end of Maud Junction station. Also constructed by the NBL during 1920, it bore the GNoSR Number 48 and later became 6848 and 2276 with the LNER. She would also be withdrawn in 1955.

Opposite top: Wednesday, 24 June 1953. The 'Princes Coronation' class of locomotives are generally viewed as being the finest of William Stanier's designs. Thirty-eight examples were constructed at Crewe Works between 1937 and 1948, of which twenty-four entered service fitted with streamlining. These streamlined locomotives were primarily used on the prestige services between London and Glasgow. The example seen here on the turntable at Ferryhill shed in Aberdeen, No. 46225 *Duchess of Gloucester*, entered service from Crewe Works in 1938 as a streamlined version. The streamlining would be removed in 1947 leaving the locomotive with a slightly sloping top to the smoke box. She would be withdrawn from service during 1964.

Opposite bottom: Wednesday, 24 June 1953. Seen here being turned manually on the Ballater turntable is ex-GER Class S69 (LNER Class B12) 4-6-0 No. 61502. Constructed at Stratford Works in 1912 she would be numbered 1502 by the GER, becoming 8502 with the LNER. Transferred to the Great Northern Section of the LNER in 1931, she would be withdrawn during 1954.

Above: Wednesday, 24 June 1953. Standing in the headshunt at Banff station yard is ex-GNoSR Class F (LNER Class D40) 4-4-0 No. 62277 *Gordon Highlander.* Numbered 49 by the GNoSR, she would become 6849 and later 2277 with the LNER. Constructed by the NBL during 1920, she was destined to join the preserved locomotives list and after withdrawal in June 1958 was repainted in GNoSR livery and subsequently ran many specials in Scotland before being housed in the Glasgow Museum of Transport.

Opposite top: Wednesday, 24 June 1953. At Inverurie Works several locomotives are awaiting scrapping. Ex-GER Class S69 (LNER Class B12) 4-6-0 No. 61528 was constructed at Stratford Works during 1914 and had been numbered 1528 by the GER. Later becoming No. 8528 with the LNER, she had been transferred to Scotland during 1933 and had been fitted with ACFI feed-water-heating apparatus during the same year. This would be removed during 1939 and the locomotive withdrawn from service in July of 1953.

Opposite bottom: Another example of the class seen here is No. 61543, which was constructed by Beardmore & Co. in Glasgow during 1920 and fitted with the ACFI equipment in 1932. She was transferred to Scotland in 1939 and the ACFI equipment was removed during 1941/42. She was withdrawn in June 1953.

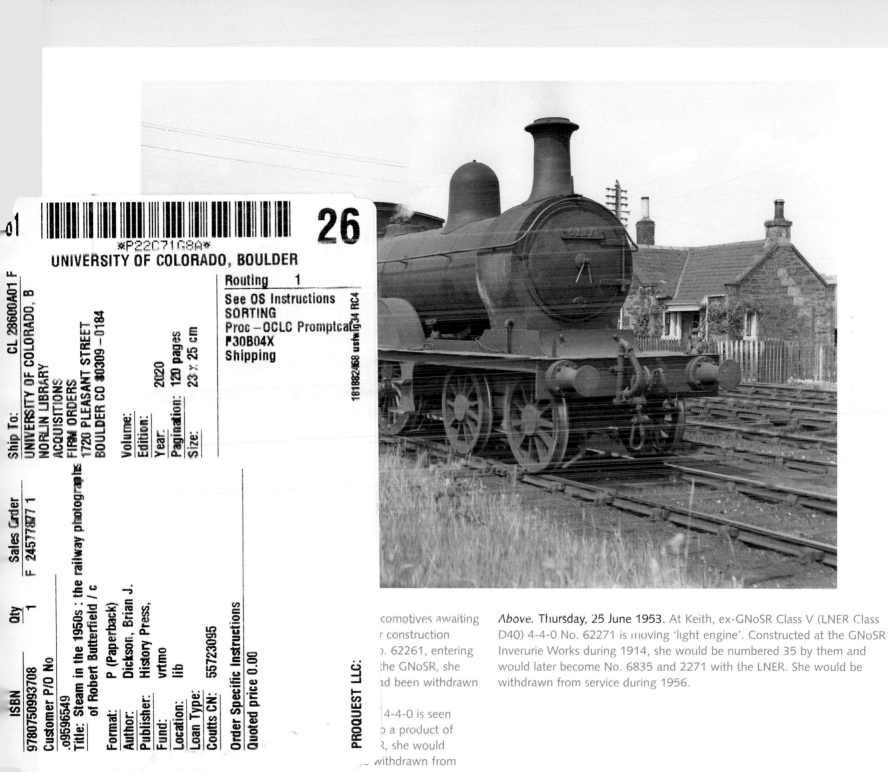

...comotives awaiting
...r construction
...o. 62261, entering
...the GNoSR, she
...ad been withdrawn

...4-4-0 is seen
...o a product of
...R, she would
...withdrawn from

service in February 1953.

Above. Thursday, 25 June 1953. At Keith, ex-GNoSR Class V (LNER Class D40) 4-4-0 No. 62271 is moving 'light engine'. Constructed at the GNoSR Inverurie Works during 1914, she would be numbered 35 by them and would later become No. 6835 and 2271 with the LNER. She would be withdrawn from service during 1956.

Thursday, 25 June 1953. At Elgin station a crew member is resting before the departure of his train to Lossiemouth. Ex-GNoSR Class V (LNER Class D40) 4-4-0 No. 62267 is seen fitted with a tender weatherboard. Constructed at Inverurie Works during 1909 and numbered 29 by the GNoSR, she would become No. 6829 and later 2267 with the LNER. She would be withdrawn in 1956.

Thursday, 25 June 1953. Seen working on shunting duties at Keith is ex-GNoSR Class V (LNER Class D40) 4-4-0 No. 62270. The last of this particular version of the class to be constructed at Inverurie Works in 1915, she would be withdrawn from service three months after this photograph in September 1953.

Thursday, 25 June 1953. Seen shunting in the yard at Elgin station is another ex-GNoSR Class V (Class D40). No. 62272 was also a product of Inverurie Works, entering service during 1910. She would be numbered 36 by the GNoSR and later become No. 6836 and 2272 with the LNER. She would be withdrawn during 1955.

Friday, 26 June 1953. Pausing between shunting duties at Inverness shed is ex-CR Class 264 (LMS Class 0F) 0-4-0 saddle tank No. 56038. The final example of the class of thirty-four locomotives to be constructed at St Rollox Works, she entered service during 1908 and would be withdrawn in 1959.

Friday, 26 June 1953. In good external condition, the Kyle of Lochalsh station pilot is a credit to the shed staff. Ex-CR Class 439 (LMS Class 2P) 0-4-4 tank No. 55216 was a product of St Rollox Works in 1912 that would be withdrawn from service during 1961. Note the horseshoe wedged into the top of the smoke box number plate.

Saturday, 27 June 1953. In the queue of bankers behind the 'down' platform at Beattock station is the unusual sight of a Pacific tank locomotive in the form of ex-CR Class 944 (LMS Class 4P) 4-6-2 tank No. 55359. Designed by William Pickersgill and intended for use on the passenger traffic between Glasgow and the Clyde Coast towns, the twelve examples of the class were delivered from the NBL during the First World War, March to May 1917. They were found more useful work during the war period and it was not until hostilities ceased that these locomotives finally did the work they were intended for. The example shown here was transferred to Beattock for banking duties around the time of Nationalisation and she was the last of her class to be withdrawn during October 1963.

Saturday, 27 June 1953. Seen here restarting a Manchester to Glasgow service from Beattock station 'down' platform having picked up a banker, is an LMS Class 6P5F Jubilee 4-6-0 No. 45712 *Victory*. Constructed at Crewe Works during 1936, she would be withdrawn in 1963. The William Stanier-designed Class 5XP or Jubilee were initially lacklustre in performance, but after early trials with an improved form of superheater these locomotives gained much better steaming qualities and an improved performance.

Above: Saturday, 27 June 1953. Another view of ex-CR Class 944 tank No. 55359 at Beattock station. Having buffered up to the rear of a goods train in the 'down' platform, she is starting to push gently.

Saturday, 27 June 1953. Two versions of the ex-CR Class 439 (LMS Class 2P) 0-4-4 tanks are seen here at Beattock shed.
Opposite top: No. 55187 constructed at St Rollox Works in 1909 during the John McIntosh period, she would be withdrawn in 1955.
Opposite bottom: No. 55232 was from the William Pickersgill period, constructed at St Rollox in 1922 and withdrawn during 1961. Note the use of a stovepipe chimney on this example.

Above: Saturday, 27 June 1953. Accelerating away from Elgin station at the head of the 1.10 p.m. working to Lossiemouth, ex-GNoSR Class V (LNER Class D40) 4-4-0 No. 62269 was a product of Inverurie Works during 1913 and numbered 28 by the GNoSR. Becoming No. 6828 and later 2269 with the LNER, she would be withdrawn in 1955. The branch service to Lossiemouth would lose its passenger services during April 1964.

Opposite top: Sunday, 5 July 1953. With Hest Bank water troughs seen in the background, ex-LMS Class 5 4-6-0 No. 44766 is passing Hest Bank station at the head of an 'up' express. Constructed at Crewe Works during 1947, she would be withdrawn twenty years later in 1967.

Opposite bottom: Sunday, 5 July 1953. Working a Glasgow to Manchester express and seen just south of Hest Bank station, ex-LMS Class 5 4-6-0 No. 45388 is piloting ex-LMS Class 6P5F Jubilee 4-6-0 No. 45700 *Amethyst*. The 'Black 5' was a product of Armstrong Whitworth & Co. in 1937 that would become one of the last steam locomotives to be withdrawn from service during August 1968. The Jubilee came into service in 1936 from Crewe Works named *Britannia*, but this would be changed in 1951 to *Amethyst* and she would be withdrawn during 1964.

Sunday, 5 July 1953. Approaching Hest Bank station with an 'up' express is ex-LMS Class 8P Princess Coronation 4-6-2 No. 46254 *City of Stoke-on-Trent*. Constructed at Crewe Works in 1946, she would be allocated to various sheds including Camden, Crewe North, Carlisle Upperby and Edge Hill, ending her days at Crewe North and being withdrawn during 1964.

Saturday, 3 October 1953. Seen approaching Mill Hill station at the head of an 'up' express to London St Pancras is ex-LMS Class 6P5F Jubilee 4-6-0 No. 45610 *Gold Coast*. Constructed at Crewe Works during 1934, she would be renamed *Ghana* in 1958 and be withdrawn from service in 1964.

Saturday, 3 October 1953. At the head of a 'down' train of milk empties from Cricklewood to Carlisle approaching Mill Hill station is ex-LMS Class 5 4-6-0 No. 45056. Constructed during 1935 by the Vulcan Foundry, she would be withdrawn in 1967.

Saturday, 3 October 1953. Pausing at Mill Hill station is a suburban passenger working from Bedford to St Pancras. In charge is BR Standard Class 4MT 2-6-4 tank No. 80046 with the fireman seen checking that the injector is operating correctly. Constructed at Brighton Works during 1952, she would initially be allocated to Bedford shed but would end her days at Corkerhill in Glasgow, being withdrawn in 1967.

Sunday, 4 October 1953. Approaching Bletchley station at the head of a Wolverhampton to London Euston express is ex-LMS Class 6P5F Jubilee 4-6-0 No. 45738 *Samson*. A product of Crewe Works in 1936, she would be withdrawn in 1963.

Sunday, 4 October 1953. Designed by Henry Fowler and introduced to the LMS during 1927, the 125 examples of his 4P 2-6-4 tank were all constructed at Derby Works. Seen here at St Albans shed is, in the foreground, No. 42341 which entered service during 1929 and would be withdrawn in 1959. In the background is No. 42335, another 1929-constructed example that would end her days at Barrow in Furness and be withdrawn during 1964.

Sunday, 4 October 1953. It is approaching 10 a.m. and waiting to depart from London Euston station at the head of the Royal Scot is ex-LMS Class 8P Princess Coronation 4-6-2 No. 46244 *King George VI*. Constructed at Crewe Works in 1940 as a streamline example of the class, the streamlining would be removed during 1947. Originally named *City of Leeds*, her identity would change during 1941 and she would be withdrawn in 1964.

Sunday, 8 November 1953. Seen here at Darlington Works is ex-WD 'Austerity', LNER Class J94 0-6-0 saddle tank No. 68008. Constructed by the Hunslet Engine Co. during 1944 she would be numbered 75101 by the War Department. Coming into LNER stock during 1946 as part of an order of seventy-five purchased by them, she would be numbered 8008 and would be withdrawn from service in 1963.

Above: Sunday, 28 March 1954. Mexborough shed lay at the heart of the South Yorkshire coalfield and consequently was allocated a large number of locomotives suitable for working heavy coal trains. This included many of the 2-8-0 wheel arrangement and seen here is ex-GCR Class 8K (LNER Class 04/8) No. 63828. Constructed by the NBL during 1918 as part of an order for the Railway Operating Division of the Royal Engineers, she would be numbered 1957 by them. Coming into LNER stock during 1928 and numbered 6609 and later 3828 by them, she would be rebuilt in 1944 with a Thompson design of standard boiler and classified Class 04/8. She would be withdrawn from service during 1965.

Opposite top: Sunday, 28 March 1954. Another ex-GCR class of locomotive is also seen here at Mexborough shed. Former GCR Class 9J (LNER Class J11) 0-6-0 No. 64432 has what appears to be coal briquettes piled on its tender. Constructed at Gorton Works during 1907 with a saturated steam boiler she would be rebuilt by the LNER during 1925 with a superheating boiler. Numbered 5318 and later 4432 by the LNER, she would be withdrawn in 1957.

Opposite bottom: Sunday, 4 April 1954. The first of the Arthur Peppercorn developments of the Class A1 Pacifics appeared from Doncaster Works in August 1948 with the last member entering service in December 1949 also from Doncaster. Of the forty-nine examples constructed, twenty-three came out of Darlington Works. Seen here passing Copley Hill in Leeds is No. 60141 *Abbotsford* at the head of the 'up' 'Harrogate Sunday Pullman' service to London King's Cross. Entering service in December 1948, she would give only sixteen years of service, being withdrawn during 1964. She was named after the residence of Sir Walter Scott situated near Melrose in the Scottish borders.

Above: Friday, 16 April 1954. A Thomas Worsdell design introduced to the NER during 1886 were the Class C (LNER Class J21) 0-6-0s. Constructed at both their Gateshead and Darlington Works, the original design was for a two-cylinder compound locomotive and 170 examples thus entered service. The remaining thirty examples were constructed as normal two-cylinder simple locomotives. All the compound versions were rebuilt over a period of years to become two-cylinder simples. Seen on the Hexham turntable is No. 65119, constructed at Gateshead Works in 1894 as a two-cylinder simple. She would be withdrawn eight months after this photograph, in December 1954.

Opposite top: Friday, 16 April 1954. Standing adjacent to Hexham shed is ex-NER Class E (LNER Class J71) 0-6-0 tank No. 68278. Designed by Thomas Worsdell and introduced during 1886, the example seen here was constructed at Darlington Works in 1892 and would give sixty-nine years of service, being withdrawn in 1961.

Opposite bottom: Friday, 16 April 1954. Thomas Worsdell's brother Wilson introduced the NER Class O (LNER Class G5) 0-4-4 tanks during 1894, all being constructed at Darlington Works with the last entering service in 1901. Designed for passenger duties they saw service throughout the North East with the last example being withdrawn from service during 1958. Standing outside Hexham shed is No. 67265, which entered service in 1896 and would be withdrawn during 1958.

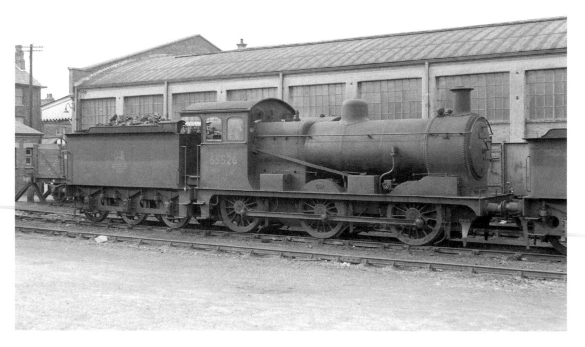

Opposite top: **Sunday, 25 April 1954.** The John Robinson design of 4-4-0 express passenger locomotives introduced during 1913 for the GCR stretched to only ten examples and were classified Class 11E by them. Becoming known as the 'Director' class, they all entered service from Gorton Works during 1913 and all were named after directors of the Great Central Railway. Seen here at Gorton shed, with a smiling crew on the footplate, is a close-up of No. 62652 *Edwin A. Beazley*. Numbered 431 by the GCR, she would become 5431 and later 2652 with the LNER who would designate them Class D10. She would be withdrawn during the following month, May 1954.

Opposite bottom: **Sunday, 9 May 1954.** At King's Lynn shed ex-GER Class F48 (LNER Class J16) 0-6-0 No. 65526 is looking in a rather grimy state. Constructed at Stratford Works in 1901 with a saturated steam boiler, she would be rebuilt with a superheating boiler during 1927 and withdrawn in 1959.

Above: **Sunday, 9 May 1954.** Concurrently with the Edward Thompson rebuilding of the six Nigel Gresley Class P2 2-8-2 locomotives into Pacifics at Doncaster during 1943 and 1944 that would be designated Class A2/2, construction of his Class A2/1 Pacifics was being carried out at Darlington Works between May 1944 and January 1945. Seen here at Doncaster shed is the third example to enter service, No. 60509 *Waverley*, which received its name in 1946. She spent her entire working life based at Haymarket shed in Edinburgh and was withdrawn having given only sixteen years of service in 1960.

Sunday, 9 May 1954. Seen here at South Lynn shed is ex-GER Class R24 (LNER Class J69) 0-6-0 tank No. 68542. Entering service from Stratford Works during 1892, she would be rebuilt on several occasions – 1905, 1924 and 1939 – before finally being seen in the form here in 1945. She would give seventy years' service and be withdrawn during 1962.

Sunday, 16 May 1954. Preparing to depart from Liverpool Street station in London at the head of an express to Norwich is BR Standard Class 7P6F 'Britannia' 4-6-2 No. 70036 *Boadicea* showing signs of a severe lack of attention. A Product of Crewe Works during 1952, she would be withdrawn from service in 1966. The 'Britannias' were the first class of the new British Railways Standard locomotives to be introduced in January 1951. A total of fifty-five examples were constructed, all at Crewe Works with the last member entering service during 1954.

Sunday, 30 May 1954. Seen here at her home shed, 5A Crewe North, is ex-LMS Class 7P Royal Scot 4-6-0 No. 46137 *The Prince of Wales's Volunteers (South Lancashire)*. Constructed by the NBL during 1927, she would originally be named *Vesta*, the new name being applied in 1936. She would be rebuilt with a taper boiler during 1955 and be withdrawn in 1962.

Sunday, 11 July 1954. The shed at Sheffield Millhouses 19B, must have been short of staff as this locomotive is in a terrible state. BR Standard Class 2 2-6-0 No. 78024 is seen at Sheffield Midland station waiting to depart at the head of a local working. Constructed at Darlington Works during 1954, she would only give eleven years' service before being withdrawn in 1965. A total of sixty-five examples of the class would all be constructed at Darlington Works, entering service between 1952 and 1956.

Tuesday, 14 September 1954. Looking a credit to the staff at Keith shed, ex-GNoSR Class F (LNER Class D 40) 4-4-0 No. 62275 *Sir David Stewart* has paused in the platform at Elgin station before moving on to some coaching stock. Constructed by the NBL in 1920, she would be withdrawn during 1955.

Sunday, 19 September 1954. Looking in ex-works condition ex-NER Class T3 (LNER Class Q7) 0-8-0 No. 63463 is standing in Tyne Dock shed yard. Designed by Vincent Raven as a powerful, superheated-boiler, three cylinder locomotive suitable to handle heavy mineral traffic, only five examples were constructed before the grouping, with a further ten examples following in 1924 all from Darlington Works. Entering service during 1919 and numbered 904 by the NER, she would become 3463 with the LNER. The whole class of fifteen locomotives was withdrawn at the end of 1962.

Sunday, 19 September 1954. Parked in Sunderland shed yard is ex-NER Class
B (LNER Class N8) 0-6-2 tank No. 69392, designed by Thomas Worsdell
and introduced during 1886 as a two-cylinder compound class to handle
heavy mineral traffic. Entering service during 1889 from Darlington Works,
this locomotive would be converted to a two-cylinder simple in 1911 and be
withdrawn from service during 1955.

Sunday, 17 October 1954. In Crewe Works yard is seen ex-LNWR Class G1 (LMS Class 6F) 0-8-0 No. 49200. Constructed at the same works in 1912, she would be withdrawn during 1959.

Sunday, 3 April 1955. Seen at Selby shed, ex-NER Class R (LNER Class D20) 4-4-0 No. 62396 is bearing a 52A Gateshead shed code. Constructed at the Gateshead Works of the NER during 1907, she would end her days working out of Alnmouth shed and be withdrawn in 1957.

Monday, 18 April 1955. At the head of what appears to be a test train – the tender is sheeted over and there are control wires appearing on the locomotive – ex-LMS Class 8P Princess Coronation 4-6-2 No. 46225 *Duchess of Gloucester* is seen near Hirst Wood, west of Saltaire. The short train consists of only four carriages.

Tuesday, 19 April 1955. Working a Bristol to Bradford express and seen near Shipley is ex-NER Class S3 (LNER Class B16) 4-6-0 No. 61410. This three-cylinder design by Vincent Raven was introduced in 1920 and seventy examples were constructed, all coming from Darlington Works. A number of the class were rebuilt by both Nigel Gresley and Edward Thompson using Walschaerts valve gear. This example would remain un-rebuilt and be withdrawn during 1960.

Sunday, 24 April 1955. Parked in the yard at Rowsley shed is ex-NLR Class 75 (LMS Class 2F) 0-6-0 tank No. 58856. Designed by J.C. Park for the North London Railway and constructed at their Bow Works, the thirty examples of the class entered service between 1879 and 1905. No. 58856 was constructed during 1896 and would be withdrawn in 1957.

Above: Sunday, 24 April 1955. Standing in Derby Works yard is newly constructed BR Standard Class 5 4-6-0 No. 73075 waiting its first allocation, which would be to Polmadie in Glasgow. Based there for its entire short working life, it would be withdrawn during 1965. The British Railways Standard '5s' were constructed by both Derby and Doncaster Works, with the first example entering service in April 1951 from Derby Works and the last appearing from Doncaster during May 1957. Of 172 locomotives, thirty were constructed utilising a Caprotti poppet valve gear.

Opposite top: Sunday, 24 April 1955. Ex-LMS Class 2P 0-4-4 tank No. 41906 is seen at Buxton. Introduced toward the end of 1932, the ten examples of this class were all constructed at Derby Works and fitted for the motor (push-pull) working of branch-line trains. She would be withdrawn during 1959.

Opposite bottom: Sunday, 24 April 1955. Seen shunting in Crewe Works yard is ex-L&YR Class 23 (LMS Class 2F) 0-6-0 saddle tank No. 51466. Originally constructed by Beyer Peacock & Co. during 1881 as an 0-6-0 tender locomotive to a design by William Barton Wright, she was rebuilt in 1896 by John Aspinall as a saddle tank. Numbered 11446 by the LMS, she would be withdrawn during 1962 having given a total of eighty-one years of service. Chalked on the side of the bunker are the words 'Atomic Flash'.

Above: Sunday, 24 April 1955. Standing adjacent to Stoke roundhouse, ex-LMS Sentinel Class four-wheeled chain-driven tank No. 47181 had been purchased by the LMS during 1930 and would be numbered 7161 by them, later becoming 7181. She would be withdrawn from service in 1956. Chalked on the side of this locomotive is the slogan 'Fish *6d* & *8d* Silver Hake'.

Opposite top: Wednesday, 11 May 1955. With the Saltaire Mills as the background, ex-LMS Class 4P 'Compound' 4-4-0 No. 41196 has just departed from Saltaire station at the head of a local working. Constructed by the Vulcan Foundry in 1927, she would be withdrawn during 1958.

Opposite bottom: Wednesday, 11 May 1955. At the head of the 'down' Thames-Clyde Express, ex-LMS Class 7P Royal Scot 4-6-0 No. 46103 *Royal Scots Fusilier* has passed Saltaire station and is seen hear Hirst Wood accelerating north. Constructed by the NBL during 1927, she would be rebuilt with a taper boiler in 1943 and be withdrawn from service in 1962.

Above: **Saturday, 21 May 1955.** At Keighley station the driver of the local to Skipton from Bradford has opened the regulator and the locomotive has almost disappeared behind the cloud of steam leaking from the glands. Ex-GNR Class N1 (LNER Class N1) 0-6-2 tank No. 69467 was the product of Doncaster Works during 1912 that would be bought by the War Office in 1914 and used as part of an armoured coastal defence gun train during the First World War. Purchased back by the LNER in 1923, she would be withdrawn from service during 1956.

Opposite top: **Sunday, 22 May 1955.** At Kingmoor shed in Carlisle, ex-LMS Class 7P Royal Scot 4-6-0 No. 46107 *Argyll and Sutherland Highlander* is parked in the yard. Constructed by the NBL during 1927, she would be rebuilt with a taper boiler in 1949 and be withdrawn in 1962.

Opposite bottom: **Sunday, 22 May 1955.** Seen at West Auckland shed bearing a 51H Kirkby Stephen shed code is Class 2MT 2-6-0 No. 46471. An example of the class introduced during 1946 by the LMS, No. 46471 was constructed post-Nationalisation in 1951 at Darlington Works. She would only give eleven years of service to be withdrawn in 1962.

Wednesday, 25 May 1955. At Hirst Wood near Saltaire, ex-LMS Class 4P 'Compound' 4-4-0 No. 41081 is in charge of a northbound mixed working that contains two six-wheeled milk tankers. Constructed at Derby Works during 1924, she would be withdrawn six months after this photograph was taken in November 1955.

Wednesday, 25 May 1955. With the fireman attempting to lay a smokescreen whilst passing Hirst Wood, ex-LMS Class 7P Royal Scot 4-6-0 No. 46133 *The King's Regiment Liverpool* is working the 'down' Thames-Clyde Express. Constructed by the NBL in 1927 and originally named *Phoenix*, she would be renamed during 1936 and be rebuilt with a taper boiler in 1944. Serving for thirty-six years, she would be withdrawn in 1963.

Tuesday, 21 June 1955. For comparison this spread shows the 2-6-2 tank locomotive designed by Nigel Gresley, introduced to the LNER in 1930 as the Class V1 primarily to work the suburban traffic around Edinburgh and Glasgow and replacing the ageing Class N2 locomotives. With a boiler pressure of 180psi the V1 class continued construction until 1939 with the last ten examples introduced utilising a boiler pressure of 200psi and being classified Class V3. This proved to be a successful introduction and subsequently, over a period between 1952 and 1961, the bulk of the V1 class were rebuilt with the higher-pressure boilers and classified V3.

Opposite top: Seen at Balloch is No. 67602, constructed at Doncaster Works in 1930, she would remain as a Class V1 and be withdrawn during 1962.
Opposite bottom: Also at Balloch is seen No. 67679, which entered service in 1939 as a Class V1 and was rebuilt as a Class V3 during 1953. She would also be withdrawn in 1962.
Above: At Helensburgh shed are seen two of the 65C Parkhead allocated locomotives utilised on the Helensburgh to Bridgeton suburban services. On the left is ex-LNER Class V3 No. 67604, which entered service as a Class V1 locomotive during 1930 and was rebuilt as a Class V3 in 1952. She would be withdrawn in 1962. On the right is ex-LNER Class V1 No. 67628, which would be rebuilt as a Class V3 during 1957 and be withdrawn in 1964.

Friday, 24 June 1955. Standing outside Wick shed is an ex-Caledonian Railway greyhound, Class 140 or 'Dunalastair IV' (LMS Class 3P) 4-4-0 No. 54439 had been constructed at St Rollox Works during 1908 as a saturated-steam locomotive and numbered 924 by the Caley. She would be rebuilt during 1915 with a superheating boiler. With 6ft 6in driving wheels, this class worked express passenger trains throughout the Caley system until replaced during LMS days by larger locomotives. She would be withdrawn in 1958.

Saturday, 25 June 1955. Designed by Peter Drummond and constructed by the Highland Railway at their Lochgorm Works during 1905 and 1906, the four diminutive examples of his 0-4-4 'Passenger Tank' Class (LMS Class 0P) locomotives found service on the lightly laid branches of that railway. No. 55051 was being utilised on the Dornoch branch and is seen here shunting at The Mound station. She would be withdrawn during 1956.

Sunday, 26 June 1955. At Kittybrewster shed, ex-NBR Class J 'Superheated Scott' (LNER Class D30) 4-4-0 No. 62421 *Laird o' Monkbarns* is looking in ex-works condition. Constructed at Cowlairs Works during 1914, she would be one of the last two examples of the class to be withdrawn from service in June 1960. The Laird o' Monkbarns character appears in the 1816-published novel *The Antiquary* by Walter Scott.

Monday, 27 June 1955. Ex-LNER Class A2 4-6-2 No. 60525 *A.H. Peppercorn* is seen here working hard at the head of the 'up Aberdonian' whist passing its home base, Ferryhill shed 61B in Aberdeen. Constructed at Darlington Works and entering service during December 1947, she was allocated to Ferryhill from 1949 and was based there for the remainder of her working life, being withdrawn during 1963.

Wednesday, 29 June 1955. This spread shows a pair of the graceful-looking ex-GNoSR Class V (LNER Class D40) 4-4-0s.

Opposite top: No. 62268 having her fire cleaned whilst standing over the ash pit at Keith shed. Note the array of darts, prickers and clinker shovels lying about.

Opposite bottom: No. 62268 having its smoke box cleared of char. A product of Inverurie Works during 1910, she would be withdrawn from service in 1956.

Above: One of the earlier constructed members of the class is seen here at Elgin shed bearing the 'Elgin No. 1' pilot disc. Note the shunter's pole wedged into the tender handrail. No. 62264 was constructed at Neilson & Co. during 1899 and would be withdrawn in 1957.

Friday, 29 June 1955. Class D40 No. 62274 *Benachie* is seen here standing inside Keith shed adjacent to ex-CR Class 439 (LMS Class 2P) 0-4-4 tank No. 55185. Constructed at St Rollox Works during 1907, No. 55185 was by this date allocated to Keith shed (61C) and would be withdrawn in 1961.

Friday, 29 June 1955. Seen standing outside Keith shed is ex-GNoSR Class F (LNER Class D40) 4-4-0 No. 62274 *Benachie.* The last locomotive to be constructed at Inverurie Works, she entered service during 1921 numbered 46 by the GNoSR, becoming 6846 and later 2274 with the LNER. She would be withdrawn from service three months later in September 1955.

Saturday, 2 July 1955. The driver of ex-LMS Class 2P 4-4-0 No. 40644 is preparing his locomotive for duty at Hurlford shed. Constructed at Crewe Works during 1931, she would only give twenty-eight years of service before being withdrawn in 1959.

Saturday, 2 July 1955. An example of the Caledonian Railway varieties of the 0-6-0 wheelbase was designed by William Pickersgill's predecessor, John McIntosh. The Class 812 was introduced during 1899 and classified 3F by the LMS. No. 57628, seen here at Ayr shed was constructed at St Rollox Works during 1899 and would give sixty-one years of service before being withdrawn in 1960. Note the shed code plate, Ayr 67C, positioned as per the LMS practice, on the upper part of the smoke box.

Sunday, 10 July 1955. Standing in Seafield shed yard, Edinburgh, attached to a four-wheeled wooden coal tender, is veteran ex-NBR Class G (LNER Class Y9) 0-4-0 saddle tank No. 68095. Constructed at Cowlairs Works during 1887 and numbered 42 by the NBR, she would carry several numbers throughout her long working life. Being withdrawn during 1962 after seventy-five years of service, she would be bought privately for preservation and is based at the SRPS site at Bo'ness.

Sunday, 10 July 1955. Parked off the turntable at South Leith shed in Edinburgh is ex-NBR Class F (LNER Class J88) 0-6-0 tank No. 68342. A product of Cowlairs Works in 1912, she would give fifty years of service, being withdrawn during 1962. This William Reid design of locomotive, primarily introduced for dock shunting duties, was first seen during 1904 with a total of thirty-five being constructed all by Cowlairs Works, the last example entering service in 1919. With a short wheelbase they were utilised in docks and yards with sharp curves such as Leith and Granton, and the last members of the class were not withdrawn until 1962.

Opposite: **Sunday, 10 July 1955.** Also seen parked in South Leith shed yard is another ex-NBR Class G (LNER Class Y9) 0-4-0 saddle tank. No. 68122 was constructed at Cowlairs Works in 1899 and would be withdrawn from service the month after this photograph was taken, August 1955.

Sunday, 10 July 1955. The Nigel Gresley Class J38 0-6-0s were specifically designed to handle the vast amounts of coal traffic in the Scottish area of the LNER, principally within Fife and around Edinburgh. The whole class was allocated to Scottish sheds: Dundee, Dunfermline, Eastfield, Thornton, St Margarets and Stirling. A total of thirty-five examples were constructed, all at Darlington Works during 1926. No. 65928 is seen here standing in the yard at Dunfermline shed. She would be one of the first of the class to be withdrawn in 1962.

Sunday, 10 July 1955. Also seen at Dunfermline shed is ex-NBR Class J 'Superheated Scott' (LNER Class D30) 4-4-0 No. 62436 *Lord Glenvarloch*. Constructed at Cowlairs Works in 1915, she would be withdrawn during 1959. The character Lord Glenvarloch appears in the 1822-published Walter Scott novel *The Fortunes of Nigel*.

Sunday, 17 July 1955. Another of the successful Nigel Gresley designs were the Class D49 4-4-0s that became known as the 'Shires' or 'Hunts'. These three-cylinder locomotives were used for secondary passenger duties primarily in Scotland and the North Eastern areas of the LNER. Construction started in 1927 with a total of seventy-six examples being delivered from Darlington Works, the last appearing during 1935. Seen at Scarborough, No. 62739 *The Badsworth* entered service in 1932 and was part of a batch fitted with Lentz rotary cam operated poppet valves. She would be withdrawn during 1960.

Sunday, 24 July 1955. Parked in the yard at Lincoln shed is ex-GER Class S56 (LNER Class J69) 0-6-0 No. 68618. Constructed at Stratford Works during 1904, she would be withdrawn in 1958.

Sunday, 24 July 1955. An earlier class of Great Eastern Railway 0-6-0 tank locomotive is seen here at Staveley shed, No. 68383 belonging to the ex-GER Class T18 (LNER Class J66). Constructed at Stratford Works in 1888, she would be withdrawn from service three months after this photograph was taken, during October 1955.

Sunday, 24 July 1955. Seen at Frodingham shed 36C is ex-GCR Class 8K (LNER Class O4) 2-8-0 No. 63747. Constructed by the NBL Queens Park Works in 1917 for the ROD and given the number 1824, she was sold to LNER in 1927 and numbered 6572 and later 3747 by that company. She would be rebuilt as Class O4/7 in 1942 and withdrawn during 1961.

Saturday, 6 August 1955. Ex-LMS Class 6P5F 'Patriot' 4-6-0 No. 45506 *The Royal Pioneer Corps* is approaching the 'up' road water trough just north of Hest Bank station at the head of a six-coach express. Constructed at Crewe Works during 1932 and given her name in 1948, she would be withdrawn after thirty years' service during 1962. A total of fifty-two examples of the class were constructed all at Crewe Works with eighteen examples being rebuilt with taper boilers between 1946 and 1949. None found their way into the preservation scene, but the 'LMS Patriot Project' is constructing No. 5551 to be named *The Unknown Warrior*.

Saturday, 6 August 1955. Passing Hest Bank station at the head of an 'up' express is ex-LMS Class 7P Royal Scot 4-6-0 No. 46138 *The London Irish Rifleman*. Constructed by the NBL in 1927, she was originally named *Fury* and was renamed in 1929. She was rebuilt with a taper boiler in 1944 and withdrawn from service during 1963.

Saturday, 6 August 1955. Approaching the Hest Bank water troughs at the head of an 'up' express is ex-LMS Class 8P Princess Coronation 4-6-2 No. 46239 *City of Chester*. Entering service from Crewe Works in 1939 as a streamlined version of the class, the streamlining being removed during 1947, she would be withdrawn in 1964.

Sunday, 11 September 1955. With Leeds City Wellington signalbox in the background, ex-LMS Class 7P Royal Scot 4-6-0 No. 46145 *The Duke of Wellington's Regiment (West Riding)* is waiting to depart with the Thames-Clyde Express. Constructed by the NBL in 1927 and originally allocated the name *Condor*, she would be rebuilt with a taper boiler in 1944 and be withdrawn during 1962.

Sunday, 18 September 1955. In sparkling ex-works condition ex-NER Class P1 (LNER Class J25) 0-6-0 No. 65663 is standing in Darlington shed yard. A total of 120 examples of the class were constructed between 1898 and 1902 at both Gateshead and Darlington Works. The example seen here was the product of Gateshead Works in 1898 and would be withdrawn after sixty-four years service during 1962.

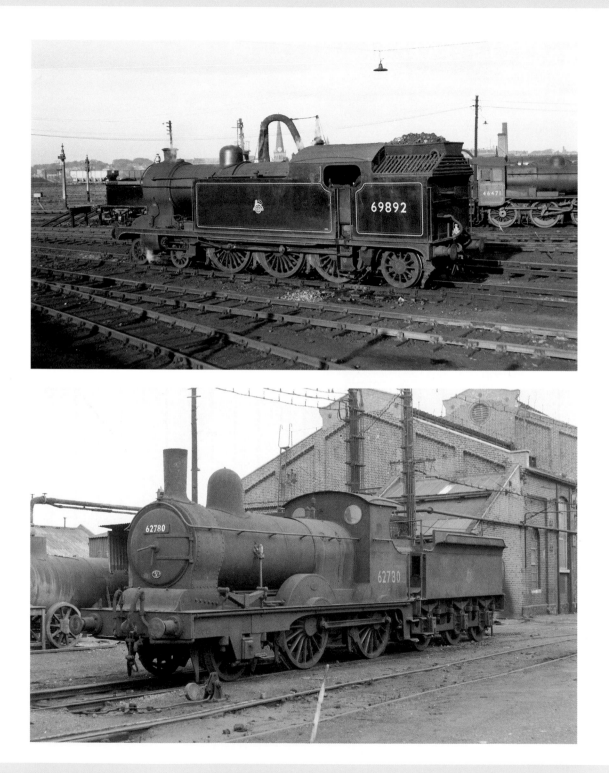

Opposite top: Sunday, 18 September 1955. Originally constructed at Darlington Works in 1922 as a three-cylinder NER Class D (LNER Class H1) 4-4-4 tank to a design by Vincent Raven, No. 69892 seen here at Darlington shed was rebuilt during 1935 as a 4-6-2 tank and classified A8 by the LNER. The 4-4-4s were found to be rather lacking in adhesion and their rebuilding to 4-6-2s led them to be more sure-footed. She would be withdrawn during 1958.

Opposite bottom: Sunday, 25 September 1955. A reminder of an earlier age, the 2-4-0 wheel arrangement of the ex-GER Class T26 (LNER Class E4) is seen here at Stratford Works with locomotive No. 62780. Constructed at the same works during 1891 to a design by James Holden, she had just been withdrawn from service during this month.

Above: Sunday, 25 September 1955. The ex-GER Class C72 (LNER Class J68) 0-6-0 tanks were a development by Arthur Hill of an earlier William Holden design. Constructed at Stratford Works in three batches of ten examples each in 1912, 1913/14 and 1923, they were utilised on both passenger traffic and shunting work. No. 68638 seen here at Stratford Works was from the 1912 batch and would be withdrawn during 1959.

Sunday, 30 October 1955. Basking in the sun at Northwich shed is ex-GCR Class 11F 'Director' (LNER Class D11) 4-4-0 No. 62662 *Prince of Wales*. Constructed at Gorton Works during 1920 and numbered 508 by the GCR, she would become 5508 and later 2662 with the LNER. She would give forty years' service to be withdrawn in 1960.

Sunday, 6 November 1955. Ex-LMS Class 2P 0-4-4 tank No. 55265 had been constructed by Nasmyth Wilson & Co. during 1925 as part of a batch of ten locomotives ordered by the LMS with the design based on the Caledonian Railway Class 439 or 'Standard Passenger' tank. Seen here at Polmadie shed, she would be withdrawn in 1962.

Opposite top: Sunday, 6 November 1955. Standing in Corkerhill shed yard is ex-LMS Class 6P5F Jubilee 4-6-0 No. 45693 *Agamemnon*. A product of Crewe Works during 1936, she would be withdrawn in 1962.

Opposite bottom: Sunday, 8 January 1956. At Trafford Park shed is seen ex-GCR Class 9H (LNER Class J10) 0-6-0 No. 65144. A Harry Pollitt development of the earlier Thomas Parker Class 9D for the GCR, she was constructed by Beyer Peacock & Co. in 1896 and would be withdrawn from service during 1958.

Above: Friday, 23 March 1956. Approaching Frizinghall station at the head of a Skipton to Bradford Forster Square local working, which consists of two coaches and one van, is ex-LNWR Class 5 (LMS Class 2P) 2-4-2 tank No. 50795. A product of Horwich Works in 1898, she would be withdrawn during 1959.

Above: Friday, 30 March 1956. This wonderful photograph shows the position of the Dillicar water troughs within the Lune Gorge with the lower reaches of Langdale Fell in the background and the Lune River adjacent to the railway. Working a fifteen coach 'up' express, ex-LMS Class 5 4-6-0 No. 45446 is seen taking water whilst piloting ex-LMS Class 8P Princess Coronation 4-6-2 No. 46243 *City of Lancaster.* The 'Black 5' was a product of Armstrong Whitworth & Co. in 1937 that would be withdrawn during 1967.

Opposite top: Sunday, 15 April 1956. With the stands of Blackpool Football Club ground at Bloomfield Road prominent in the Background, ex-LMS Class 4P 'Compound' 4-4-0 No. 41102 is 'out of steam'. Constructed at Derby Works during 1924, she would be withdrawn from service in 1958.

Opposite bottom: Sunday, 29 April 1956. Designed by Henry Fowler for the Midland Railway specifically to handle goods and mineral traffic on the Somerset and Dorset Joint Railway, the Class 7F 2-8-0s were highly successful in that task. Seen here at Derby Works is No. 53807, which was constructed by Robert Stephenson & Co. during 1925. She was numbered 87 by the S&DJR, later becoming 9677 and 13897 by the LMS. She would be withdrawn from service in 1964.

Sunday, 20 May 1956. With excess water draining from the tender, ex-LMS Class 5 4-6-0 No. 45196 is seen at the head of an 'up' local working as it approaches Hest Bank station. A product of Armstrong Whitworth & Co. during 1935, she would be withdrawn in 1967.

Sunday, 20 May 1956. Seen approaching the water troughs to the north of Hest Bank station the 'up' Royal Scot is in the hands of ex-LMS Class 8P Princess Coronation 4-6-2 No. 46229 *Duchess of Hamilton.* Constructed at Crewe Works during 1938 and entering service in a streamlined form that would be removed in 1947, she would be withdrawn from service in 1964. The authorities running the 1939 World Fair in America invited the LMS to send a locomotive and train for display and the LMS chose No. 6220 *Coronation*

to represent them. With No. 6229 being in better condition it was decided to exchange identities for the visit, numbers and names reverting on return of the locomotive during 1942. Purchased privately, she was put on display at the Butlins Holiday camp at Minehead. Removed from there during 1975, she was later returned to steam and would haul many specials during the 1980s and 1990s. She has since been on static display at the National Railway Museum in York and since 2009 has been seen with her streamlining reinstated.

Sunday, 27 May 1956. Seen in New England shed is ex-GNR Class C2 (LNER Class C12) 4-4-2 tank No. 67376. A member of the first class of tank engines introduced by Henry Ivatt during 1898, the example seen here entered service from Doncaster Works in 1901 fitted with condensing gear to work the London suburban services. Numbered 1521 by the GNR, she would become 4521 and 7376 with the LNER. The condensing gear would be removed during 1923 and she would be withdrawn from service in 1958.

Sunday, 27 May 1956. Also in a New England shed siding, 'out of steam', is ex-GER Class S69 (LNER Class B12) 4-6-0 No. 61567. Constructed at Stratford Works during 1920, she would be numbered 1567 by the GER and 8567 by the LNER. Fitted with ACFI Feed Water Apparatus during 1932, this would be removed in 1935. Allocated to several Eastern Region sheds, she would finally be based at Cambridge before being withdrawn in 1958.

Sunday, 27 May 1956. Designed by Thomas Parker for the Great Central Railway, his Class 9F 0-6-2 tanks were introduced during 1893 with thirty examples being constructed by Beyer Peacock & Co. in 1893 and 1894. Over 120 examples appeared with the last entering service in 1901. Primarily intended to handle goods traffic, they were eventually found allocated throughout former LNER territory. Classified N5 by the LNER, No. 69277 is seen here at Newark shed, entering service in 1894 from Beyer Peacock & Co. she would be numbered 544 by the GCR and later 5544 and 9277 with the LNER. She would be withdrawn six months after this photograph in November 1956.

Sunday, 27 May 1956. Introduced during 1891 for the Great Eastern Railway, the James Holden design of 2-4-0 mixed-traffic locomotives were classified T26 by them. Becoming Class E4 with the LNER, No. 62786 seen here at Cambridge shed was constructed at Stratford Works during 1891 and would be withdrawn two months after this photograph in July 1956.

Sunday, 27 May 1956. Preparing to depart from March shed is ex-GER Class S46 (LNER Class D16) 4-4-0 No. 62529. Constructed at Stratford Works during 1902, she would be rebuilt with a superheating boiler in 1931 and be classified Class D15 by the LNER. Being further rebuilt in 1935, she would be classified Class D16. Having given fifty-seven years of service she would be withdrawn during 1959.

Friday, 20 July 1956. With a train of cattle wagons in the rear, ex-LMS Class 6P5F (later Class 5) 2-6-0 No. 42971 is seen working in the 'down' direction at Colwyn Bay station. A product of Crewe Works during 1934, she would be numbered 13271 and later 2971 by the LMS. Based at Mold Junction (6B) at this date, she would be withdrawn after thirty years' service in 1964. Nominally credited to William Stanier as the designer, the class of forty examples were constructed at Crewe Works.

Wednesday, 25 July 1956. Having been withdrawn from service earlier in the same month, ex-GER Class M15 (LNER Class F4) 2-4-2 tank No. 67157 is seen here at Kittybrewster shed awaiting removal for scrapping. Introduced by Thomas Worsdell during 1884, this example was constructed at Stratford Works in 1907 and was sent north in 1948 to work the St Combs branch from Fraserburgh. With a very light axle weight of just less than 15 tons, she was one of only four members of the class sent to Scotland for this work. Her last duties were as Works pilot at Inverurie.

Wednesday, 25 July 1956. Purchased from Manning Wardle & Co. during 1915, this ex-GNoSR Class X (LNER Class Z4) 0-4-2 tank No. 68191, seen here in the harbour area, was one of a pair of examples purchased for shunting around Aberdeen Harbour. Initially numbered 117 by the GNoSR, she soon became No. 44 with them and with the LNER became No. 6844 and later 8191. She would be withdrawn during 1959.

Tuesday, 2 April 1957. Seen near Clifton south of Penrith, ex-LMS Class 5 4-6-0 No. 45310 is at the head of a long goods train. Constructed in 1937 by Armstrong Whitworth & Co. she would be allocated throughout her working life to sheds as far apart as Willesden and Carnforth and would be withdrawn during August 1968.

Friday, 5 April 1957. The first few carriages of this eleven coach 'up' express are receiving a soaking as ex-LMS Class 7P Royal Scot 4-6-0 No. 46126 *Royal Army Service Corps* passes over Dillicar water troughs just south of Tebay. Constructed by the NBL in Glasgow during 1927 and originally named *Sans Pareil*, she would be renamed in 1936 and be rebuilt with a taper boiler during 1945. Something of a wanderer, she was allocated to sheds as far apart as Carlisle Upperby and Willesden, being withdrawn in 1963.

Friday, 5 April 1957. With a crew member handing over the single-line token to the signalman at Coniston, Class 2P 2-6-2 tank No. 41221 is entering the branch terminus. Introduced by Henry George Ivatt during 1946, building of the class continued until 1952 by which time 130 examples had entered service. Constructed at both Derby and Crewe Works, No. 41221 was a product of Crewe during 1948 and would only give seventeen years service before being withdrawn in 1965. This former Furness Railway branch from its junction at Foxfield would succumb to closure eighteen months later in October 1958.

Monday, 8 April 1957. Waiting to depart from Carlisle Citadel station at the head of a local working to Edinburgh via Hawick, the Waverley Route, is ex-NBR Class J 'Superheated Scott' (LNER Class D30) 4-4-0 No. 62423 *Dugald Dalgetty*. The class consisted of twenty-seven examples all constructed at Cowlairs Works and named after characters from the novels of Sir Walter Scott. No. 62423 entered service in 1914 numbered 414 by the NBR, becoming 9414 and later 2423 with the LNER. She would be withdrawn from service during December 1957. The character Dugald Dalgetty is met in the 1819-published Scott novel *A Legend of Montrose*.

Monday, 8 April 1957. Pausing between duties as Carlisle Citadel station pilot, ex-LMS Class 3F 0-6-0 No. 47332 was a product of the NBL in Glasgow in 1926. She would be withdrawn during 1962. Constructed by no fewer than five outside manufacturers in addition to the LMS Works at Horwich, a total of 422 appeared between 1924 and 1931. W.G. Bagnall Ltd, William Beardmore & Co., the Hunslet Engine Co., the North British Locomotive Co. and the Vulcan Foundry contributed a total of 407 examples.

Monday, 8 April 1957. Accelerating away from Carlisle Citadel station at the head of a Carlisle to Edinburgh express via the Waverley Route is Class A2 4-6-2 No. 60537 *Batchelor's Button*. Constructed at Doncaster Works during 1948 and named after the 1905 Doncaster Cup race winner, she would only give fourteen years' service before being withdrawn in 1962.

Sunday, 19 May 1957. At Manningham shed a pair of ex-L&YR Class 5 (LMS Class 2P) 2-4-2 tanks are seen in store. Designed by John Aspinall and introduced during 1889, the class extended to 270 examples with the last appearing in 1901, all constructed at Horwich Works. No. 50636 entered service in 1890 and would be numbered 1039 by the L&YR becoming 10636 with the LMS. She was withdrawn during the month of this photograph.

No. 50795 was constructed in 1898 and numbered 1378 by the L&YR and 10795 by the LMS. She would survive a further two years before being withdrawn in 1959.

Sunday, 1 September 1957. On the approach to Hest Bank station is ex-LMS Class 7P 'Rebuilt Patriot' 4-6-0 No. 45531 *Sir Frederick Harrison* at the head of an 'up' milk train. The 'Patriot' class of 3-cylinder 4-6-0's were originally conceived by Henry Fowler as rebuilds of the ex-LNWR 'Claughton' four-cylinder class with a total of fifty-two examples being constructed. Entering service from Crewe Works during 1933 and numbered 5531 by the LMS, she would be named in 1937, the name being transferred from classmate No. 5524. She would be rebuilt with a taper boiler during 1947 and be withdrawn from service in 1965.

Sunday, 1 September 1957. Seen passing Hest Bank station, the 'up' Royal Scot is in the hands of ex-LMS Class 7P 'Rebuilt Jubilee' 4-6-0 No. 45735 *Comet*. Originally constructed at Crewe Works as a Jubilee-class locomotive during 1936, she would be rebuilt in 1942 with a larger boiler and firegrate, the increased boiler pressure and greater superheating leading to increased tractive effort. She was also fitted with a double chimney and later with smoke deflector plates. She would be withdrawn from service in 1964.

Sunday, 1 September 1957. At Hest Bank station, ex-LMS Class 8P Princess
Coronation 4-6-2 No. 46256 *Sir William A. Stanier F.R.S.* is seen at the head
of a Glasgow to London Euston working. Entering service at the end of 1947
from Crewe Works, she had detail differences introduced by Henry George
Ivatt such as roller bearings and a revised trailing axle. Destined to spend only
seventeen years in service, she would be withdrawn during 1964.

Sunday, 8 September 1957. Near Pilmoor, north of York on the East Coast Main Line, the photographer was able to capture some of the express workings on that day.

Opposite top: Ex-LNER Class A3 4-6-2 No. 60080 *Dick Turpin* was constructed by the NBL at their Hyde Park Works during 1924 as a Class A1 locomotive and originally numbered 2579 by the LNER. Rebuilt as a Class A3 locomotive during 1942, she would have a double chimney fitted In 1959 and trough-style smoke deflectors fitted during 1961. She is seen here working the 'down' 'Northumbrian' express from London King's Cross to Newcastle. Allocated to sheds in the Newcastle and Leeds areas throughout her working life, she would be withdrawn in 1964.

Opposite bottom: Another ex-LNER Class A3 4-6-2 is working an 'up' Edinburgh to London King's Cross express. No. 60048 *Doncaster* was constructed at the works of the same name during 1924 as a Class A1 locomotive and would be rebuilt as a Class A3 in 1946. She would have a double chimney fitted during 1959 and smoke deflectors fitted in 1961. Originally numbered 2547 by the LNER, she was named after the 1873 Derby winner and would be variously allocated to King's Cross, Grantham, Neasden and Doncaster sheds throughout her working life, being withdrawn in 1963.

Above: The final members of the Class A2 series of locomotives were the Arthur Peppercorn developments with the last examples entering service during August 1948 from Doncaster Works. No. 60539 *Bronzino* is seen in charge of a 'down' London King's Cross to Glasgow express. Named after the winner of the 1910 Doncaster Cup, she would only give fourteen years' service before withdrawal in 1962.

Sunday, 19 July 1959. Working an 'up' express near the site of the former Brock water troughs is ex-LMS Class 8P Princess Coronation 4-6-2 No. 46243 *City of Lancaster*. Constructed during 1940 at Crewe Works as a streamlined version, the streamlining being removed in 1949, she would be withdrawn during 1964.